Life by a Bay

by Sara E. Turner

Contents

Science Vocabulary

life cycle

A **life cycle** is the way a living thing grows, changes, makes more living things like itself, and dies.

A pine tree grows and changes during its **life cycle.**

germinate

When seeds **germinate,** they begin to grow.

These pine tree seeds have **germinated,** or begun to grow.

seedling

A **seedling** is a young plant.

This **seedling** is a very young pine tree.

amphibian

An **amphibian** starts its life in water and moves to land.

bird

A **bird** has feathers.

fish

A **fish** lives in water and has gills.

insect

An **insect** has three pairs of legs.

mammal

A **mammal** is born live.

reptile

A **reptile** has scaly skin.

extinct

A living thing becomes **extinct** when no members of its group are alive.

SuperCroc is **extinct.**

My Science Vocabulary

- amphibian
- bird
- extinct
- fish
- germinate
- insect
- life cycle
- mammal
- reptile
- seedling

Plants by a Bay

Many kinds of plants grow by a Florida bay. A bay is a body of water.

Slash pine trees grow by a Florida bay. Like all plants, they have a **life cycle**.

Land surrounds parts of a bay. This land has a lot of slash pine trees.

life cycle

A **life cycle** is the way a living thing grows, changes, makes more living things like itself, and dies.

A slash pine begins its life cycle as a seed.
Slash pines have cones that form seeds.

A seed can begin to grow, or **germinate.**
The **seedling** grows quickly in its first year.

germinate

When seeds **germinate,** they begin to grow.

seedling

A **seedling** is a young plant.

The seedling becomes a young tree. The young tree grows taller and has needle-like leaves.

The adult slash pine makes cones every four years. When seeds fall out of the cones, new slash pines can begin to grow.

Some slash pine trees can live as long as 200 years.

A Slash Pine Life Cycle

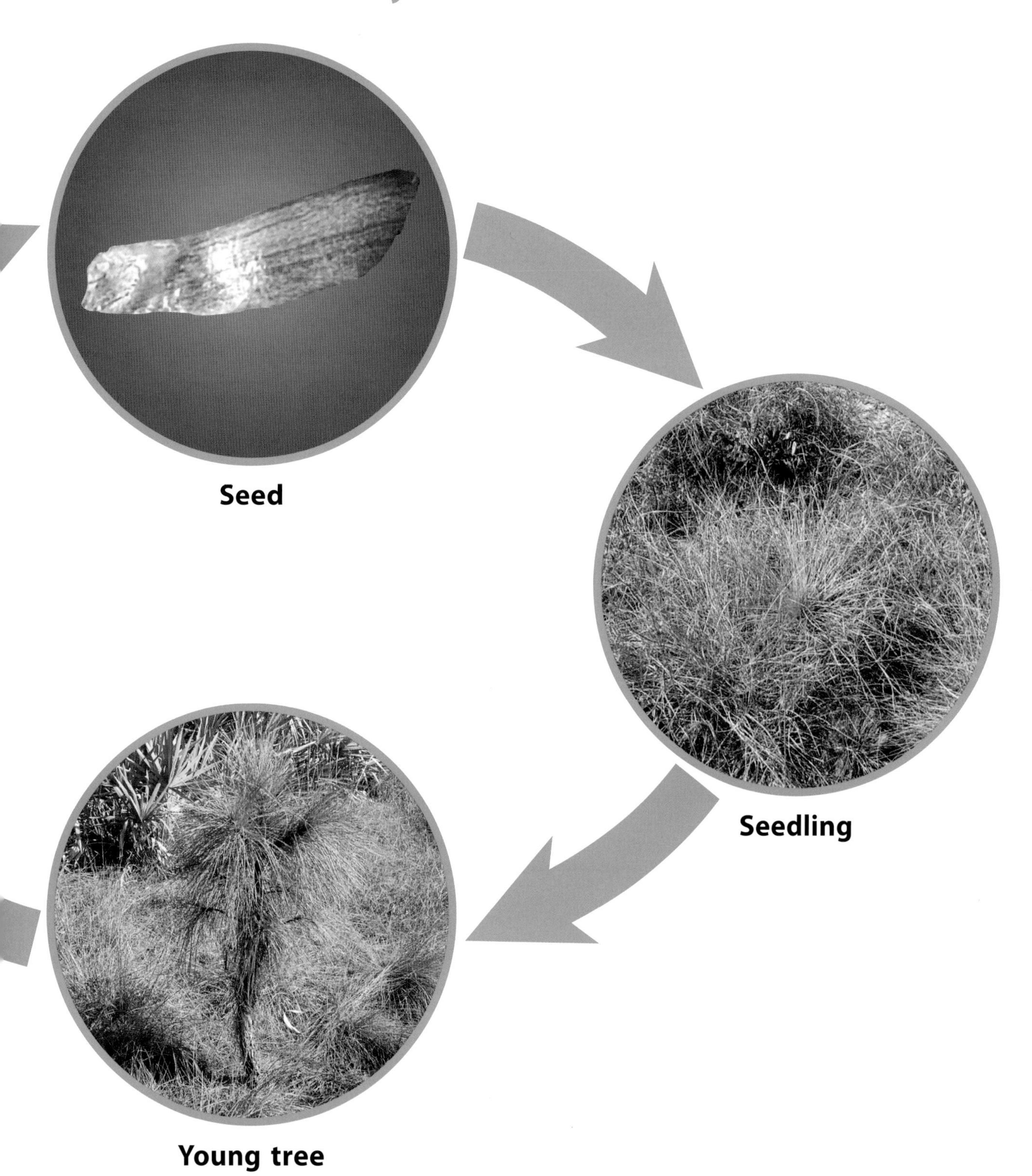

Animals by a Bay

Animals in and near a Florida bay, such as dolphins and pelicans, grow and change, too. Dolphins are **mammals**.

mammal

A **mammal** has a backbone, usually has hair or fur, and its young are born live.

Pelicans are **birds.** Baby pelicans look like their parents, but they are a different color.

Pelicans

bird

A **bird** has feathers, two legs, two wings, and lays eggs.

Butterflies and **fish** live by a bay. Butterflies are **insects.** Insects hatch from eggs. Most fish do, too.

A butterfly has three body parts and six legs.

A damselfish lives in water and has gills.

fish

A **fish** has a backbone, lives in water, and has gills.

insect

An **insect** has three body parts, six legs, and hatches from an egg.

Tadpoles also hatch from eggs. They grow to become frogs. Frogs are **amphibians.**

A frog lives in water and on land.

amphibian

An **amphibian** has a backbone and lives part of its life in water and on land.

A crocodile is a **reptile.** A crocodile also begins its life cycle as an egg.

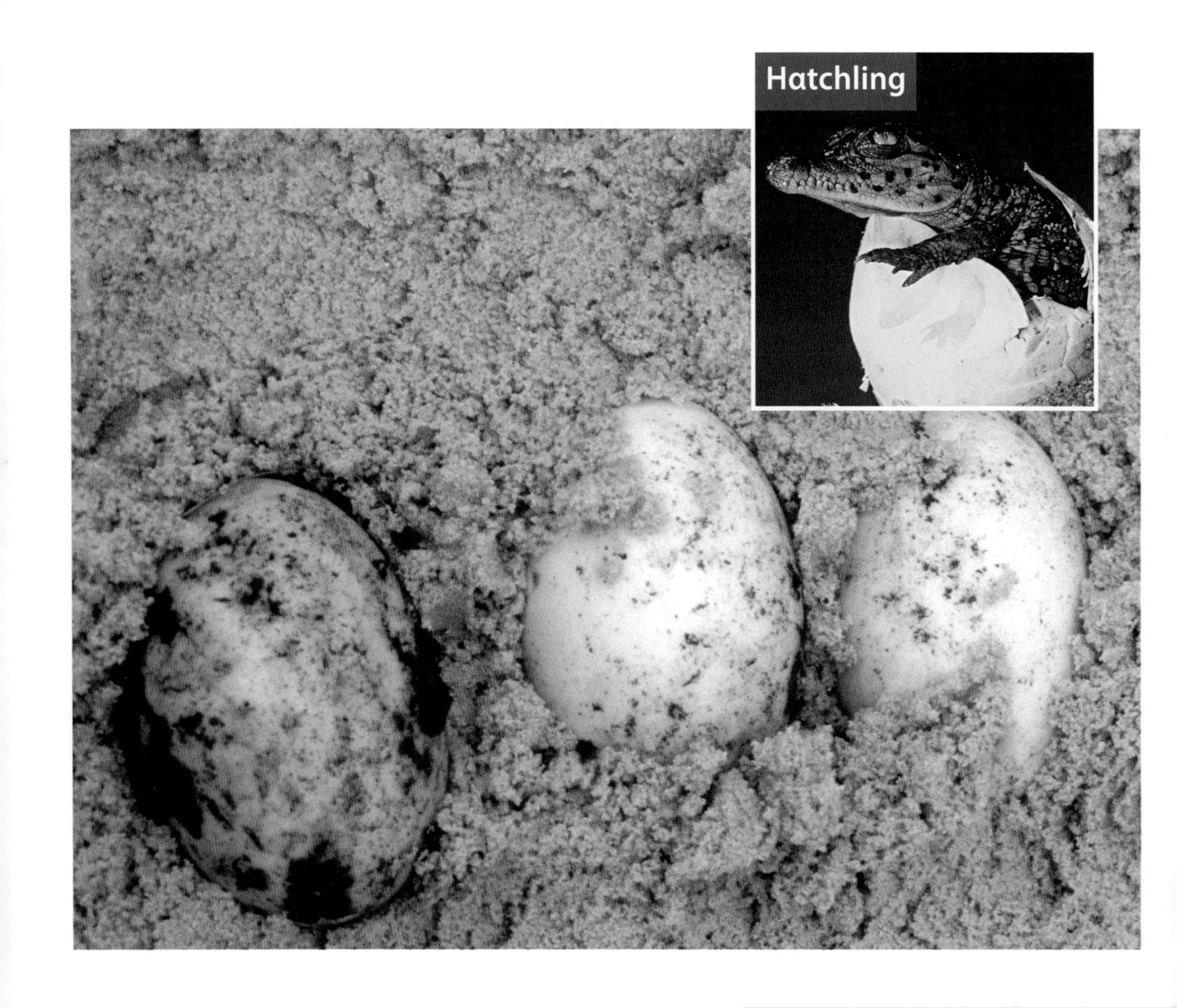

reptile

A **reptile** has a backbone, scaly skin, and breathes with lungs.

A young crocodile has dark stripes on its back. An adult crocodile doesn't have stripes.

Like all animals, a crocodile has a life cycle. How does a crocodile grow and change?

An American Crocodile Life Cycle

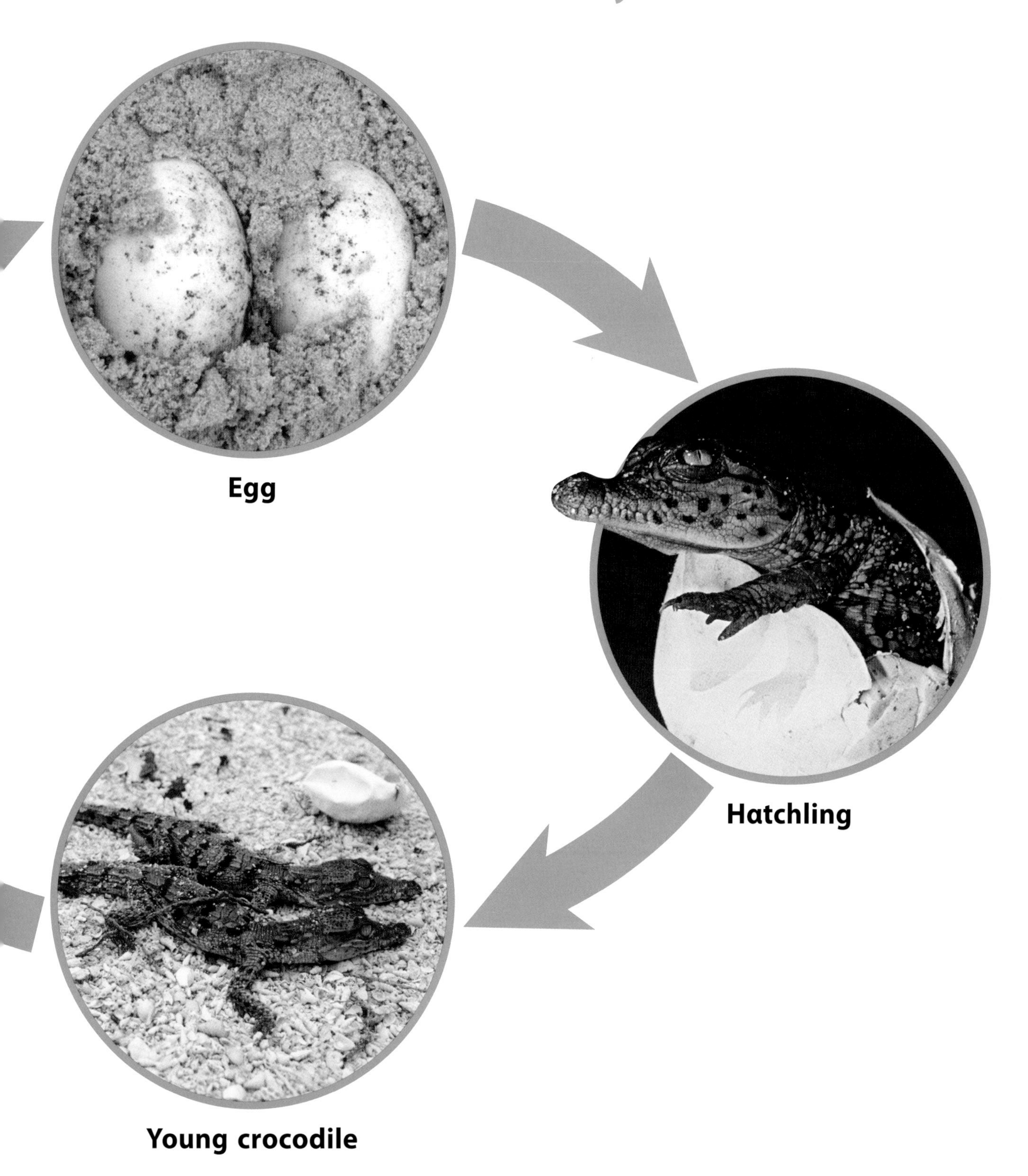

The Crocodile That Became Extinct

SuperCroc is a crocodile that lived long ago. Then this kind of crocodile died out, or became **extinct.**

SuperCroc lived during the time of the dinosaurs.

extinct

A living thing becomes **extinct** when no members of its group are alive.

Scientists don't know why. For some reason, these crocodiles could not begin new life cycles.

Crocodiles live in many different places on Earth. American crocodiles grow and change by a Florida bay.

Along with plants and other animals they begin and end their life cycles by a bay.

Conclusion

American crocodiles and slash pines can begin their life cycles by a Florida bay. Some plants and animals cannot begin new life cycles. SuperCroc could not and became extinct.

Think About the Big Ideas

1. How do slash pine trees grow and change?
2. How do American crocodiles grow and change?
3. How do scientists think SuperCroc became extinct?

Share and Compare

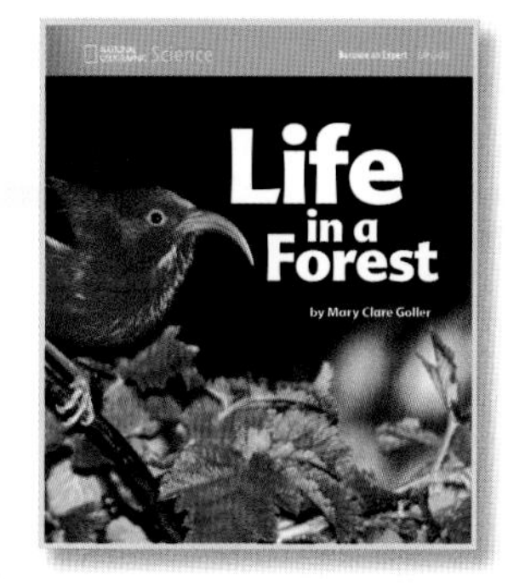

Turn and Talk

Compare the life cycles in your books. How are they alike? How are they different?

Read

Find a photo with a caption and read it to a classmate.

Write

Describe the plants and animals in your book. Share what you wrote with a classmate.

Draw

Draw one stage of the life cycle of a plant or animal from your book. Share your drawing.

Meet Mireya Mayor

Scientists observe plants and animals. They record their observations.

Mireya Mayor is a scientist who studies plants and animals. She looks for patterns, such as the way living things grow and change. She writes notes about these patterns.

Mireya has studied many animals. She takes notes as she observes them in the wild.

Mireya Mayor holds a silky sifaka.

Index

Acknowledgments
Grateful acknowledgment is given to the authors, artists, photographers, museums, publishers, and agents for permission to reprint copyrighted material. Every effort has been made to secure the appropriate permission. If any omissions have been made or if corrections are required, please contact the Publisher.

Photographic Credits:
Cover (bg) Otis Imboden/National Geographic Image Collection; Cvr Flap (t), 4 (cr), 5 (b), 11 (bg) Photo courtesy of the Florida Division of Forestry; Cvr Flap (c), 6 (tr), 17 Annie Griffiths Belt/National Geographic Image Collection; Cvr Flap (b), 7 (b), 24 Reuters/Corbis; Title (bg) Chris Johns/National Geographic Image Collection; 2-3 Steve Byland/iStockphoto; 4 (t, b, cl), 15 (t, c, b) Photo courtesy of the Florida Division of Forestry; 6 (tl), 19 (bg) Digital Vision/Getty Images; b (bl), 18 (r) Doug Perrine/SeaPics.com; 6 (r), 18 (l) Creatas/Jupiterimages; 7 (tl), 16 Konrad Wothe/Animals Animals; 7 (tr), 28 Joe Mc Donald/Animals Animals; 8 (l) Patti Murray/Animals Animals, (r) Connie Bransilver/Photo Researchers, Inc.; 9 Raymond Gehan/National Geographic Image Collection; 10 (bg) David Hosking/Frank Lane Picture Agency/Corbis, (inset) Photo courtesy of the Florida Division of Forestry; 12 Photo courtesy of the Florida Division of Forestry; 13 (bg), 14 (1, c) Photo courtesy of the Florida Division of Forestry, 13 (inset), 14 (r) F. Poelking/Arco Images GmbH/Alamy Images; 19 (inset) Jean Hall; Cordaiy Photo Library Ltd./Corbis; 20 (bg), 23(t) Ted Levin/Animals Animals, 20 (inset), 23 (c) Miller/Stock Image/Jupiterimages; 21, 23 (b) Jim Doran/Animals Animals; 22 (l) Lynn M. Stone/Nature Picture Library/Alamy Images, (r), 26 (inset) Millard H. Sharp/Photo Researchers, Inc.; 25 IOS Bios - Auteurs (droits gérés) Gunther Michel/Peter Arnold, Inc.; 26-27 (bg) Raul Touzon/National Geographic Image Collection; 30-31 Mark Thiessen/National Geographic Image Collection; Inside Back Cover (bg) Romilly Lockyer/Getty Images.

Illustrations:
Paul Mirocha

Published by National Geographic School Publishing & Hampton-Brown
Sheron Long, Chairman
Samuel Gesumaria, Vice-Chairman
Alison Wagner, President and CEO
Susan Schaffrath, Executive Vice President, Product Development

Editorial: Fawn Bailey, Joseph Baron, Carl Benoit, Jennifer Cocson, Francis Downey, Richard Easby, Mary Clare Goller, Chris Jaeggi, Carol Kotlarczyk, Kathleen Lally, Henry Layne, Allison Lim, Taunya Nesin, Paul Osborn, Chris Siegel, Sara Turner, Lara Winegar, Barbara Wood

Art, Design, and Production: Andrea Cockrum, Kim Cockrum, Adriana Cordero, Darius Detwiler, Alicia DiPiero, David Dumo, Jean Elam, Jeri Gibson, Shanin Glenn, Raymond Godfrey, Raymond Hoffmeyer, Rick Holcomb, Cynthia Lee, Anna Matras, Gordon McAlpin, Melina Meltzer, Rick Morrison, Cindy Olson, Christiana Overman, Andrea Pastrano-Tamez, Sean Philpotts, Leonard Pierce, Cathy Revers, Stephanie Rice, Christopher Roy, Janet Sandbach, Susan Scheuer, Margaret Sidlosky, Jonni Stains, Shane Tackett, Andrea Thompson, Andrea Troxel, Ana Vela, Teri Wilson, Brown Publishing Network, Chaos Factory, Inc., Feldman and Associates, Inc.

The National Geographic Society
John M. Fahey, Jr., President & Chief Executive Officer
Gilbert M. Grosvenor, Chairman of the Board

Manufacturing and Quality Management, The National Geographic Society
Christoper A. Liedel, Chief Financial Officer
George Bounelis, Vice President

National Geographic School Publishing
Hampton-Brown
P.O. Box 223220
Carmel, California 93922
www.NGSP.com

Printed in the USA.

ISBN: 978-0-7362-5568-4

10 11 12 13 14 15 16 17

10 9 8 7 6 5 4 3 2 1